Detectives: Animal Smarts

Search for the Facts...

Keeping in Touch

Anne O'Daly

BROWN BEAR BOOKS

Published by Brown Bear Books Ltd
4877 N. Circulo Bujia
Tucson, AZ 85718
USA

and

G14, Regent Studios
1 Thane Villas
London N7 7PH
UK

ISBN 978-1-83572-026-4 (ALB)
ISBN 978-1-83572-032-5 (paperback)
ISBN 978-1-83572-038-7 (ebook)

Library of Congress Cataloging-in-Publication Data available on request

Design Manager: Keith Davis
Children's Publisher: Anne O'Daly
Picture Manager: Sophie Mortimer

Picture Credits
Cover: iStock: 1001slide. Interior: iStock: Byrdyak 12-13, Gerald Corsi 18-10, rebius 1, 5b, skynesher 4b; Shutterstock: ArCaLu 21, Bildagentur Zoonar GmbH 18, Chased'annumulls 4t, Earth Theater 6-7, Sue Green 8-9, Fer Gregory 5tr, Harris Motion Photo 22bl, Ingrid Heres 22tl, KH Lung Center 10-11, Oleksii Lomako 14-15, Jason Mintzer 16-17, MPRshots 6, 23, Brian Stuart el 13, Mike Norkum 4-5b, JONATHAN PLEDGER 5tl, Tono Pulido 22tr, Mark Schooken 20, Debbie Steinhausser 4-5c, 22br, Johan Swanepoel 8, PaxPhong Trinh 14, UWPhotog 5c, Wood Photography LLC 16.
t=top, b=bottom, l=left, r=right, c=center
All artwork and other photography Brown Bear Books.

Brown Bear Books has made every attempt to contact the copyright holder.
If you have any information about omissions, please contact: licensing@brownbearbooks.co.uk

Manufactured in the United States of America
CPSIA compliance information: Batch#AG/5663

Websites
The website addresses in this book were valid at the time of going to press. However, it is possible that contents or addresses may change following publication of this book. No responsibility for any such changes can be accepted by the author or the publisher. Readers should be supervised when they access the Internet.

Contents

Meet the Communicators 4

Fact Files

- Humpback Whale 6
- Elephant 8
- Firefly 10
- Lion 12
- Peacock 14
- Rattlesnake 16
- Skunk 18
- Woodpecker 20

Quiz 22

Useful Words 23

Find Out More 24

Index 24

Meet the Communicators

Animals send messages.
Some use special sounds.
Others flash light signals.
Find out how animals keep in touch!

Keeping In Touch

Mother and baby whales call to each other. They use quiet sounds. Predators can't hear them.

Humpback Whale

Humpback whales sing to keep in touch. The songs travel through the ocean. The whales can sing for five hours.

FACT FILE

Name: humpback whale

Where it lives: oceans around the world

What it eats: small fish and krill

Type of animal: mammal

Size: 38 to 50 feet (11.5 to 15 m)

Body: long flippers and strong tail for swimming

A humpback whale jumps from the water.

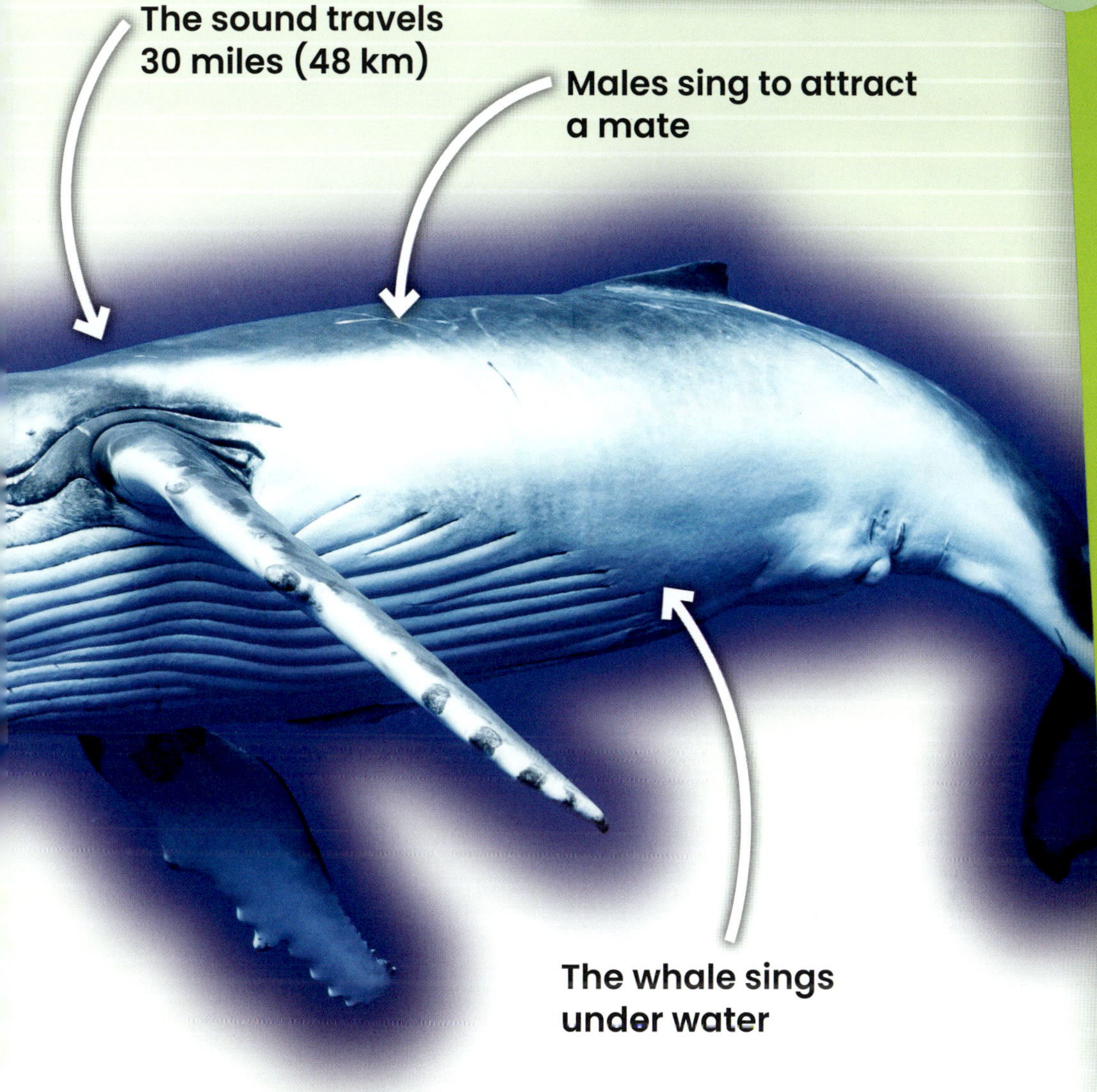

MINI FACTS

Whales migrate. They have their babies in warm water. They swim to cold waters to feed. The whales travel thousands of miles each year.

Elephant

Elephants make low rumbling sounds. These noises are called infrasound. They are too low for people to hear. Elephants can hear them from miles away.

FACT FILE

Name: African elephant

Where it lives: grasslands and woodlands in Africa

What it eats: grasses, leaves, fruits, and roots

Type of animal: mammal

Size: up to 11 feet (3.4 m)

Body: gray color, large ears, long trunk

Elephants flap their ears when they are excited

Elephants communicate with touch. They use their trunk to stroke each other.

Elephants make trumpet sounds with their trunks

MINI FACTS

Elephants use smell to find their way. They follow the scent of poop left by other elephants.

They pick up low noises through their feet

Firefly

Fireflies send messages with light. These beetles come out at night. They flash to send signals in the dark.

FACT FILE

Name: firefly (sometimes called lightning bug)

Where it lives: parks, gardens, and woods

What it eats: nectar and pollen; some eat other fireflies

Type of animal: insect

Size: up to 1 inch (2.5 cm)

Body: brown or black with red or yellow marks

Keeping In Touch

Thousands of fireflies light up a forest.

MINI FACTS

Most fireflies taste bad. Predators leave them alone. The light warns predators to stay away.

Each kind of firefly has its own pattern of flashes

Lion

Lions live together in groups.
A group of lions is a pride.
Lions have ways to keep in touch.
They use sound, touch, and scents.

FACT FILE

Name: lion

Where it lives: grasslands in Africa; there are some lions in India

What it eats: other animals

Type of animal: mammal

Size: males up 82 inches (2 m), females up to 72 inches (1.8 m)

Body: males have a shaggy mane around their head

A male roars to warn other lions

Lions growl and purr to show they are happy

MINI FACTS

A lion's roar is loud. It can be heard from 5 miles (8 km) away!

They nuzzle and rub heads to say hello

Males make a strong scent. They spray trees and rocks. The smell warns other lions to stay away.

Peacock

Peacocks are famous for their tails. Only the male has colorful feathers. He uses it to attract a mate.

FACT FILE

Name: Indian peacock

Where it lives: forests and fields

What it eats: plants, insects, and other small animals

Type of animal: bird

Size: males up to 98 inches (2.5 m) long, females up to 43 inches (1 m) long

Body: males have a colorful tail like a huge fan

The males are called peacocks. Females are peahens. They have brown feathers.

MINI FACTS

Peacocks lose their tail feathers each year. This is molting. New feathers grow in about seven months.

Rattlesnake

Rattlesnakes have a rattle on their tail. If a snake is in danger, it shakes the rattle. The noise says "stay away!"

FACT FILE

Name: red diamond rattlesnake

Where it lives: Southwest California and Baja California

What it eats: lizards, birds, and small mammals

Type of animal: reptile

Size: 5 feet (1.5 m)

Body: head and body are pink or red; diamond shapes on the back

Babies are born with a button on their tail. New scales grow as the snakes get older.

MINI FACTS

The rattle is made of keratin. Our nails are made of keratin, too.

Skunk

When a skunk is alarmed, it stamps its feet. If that doesn't work, it does something else. The skunk lifts its tail. It squirts a stinky liquid.

FACT FILE

Name: skunk

Where it lives: forests, deserts, and mountains

What it eats: plants, insects, lizards, and eggs

Type of animal: mammal

Size: up to 37 inches (94 cm) long

Body: black fur with white stripes

Black and white patterns warn predators to keep away

Baby skunks are called kits. They can spray from about three months old.

MINI FACTS

A skunk aims the spray at its enemy's eyes. The enemy can't see. The skunk can run away.

Woodpecker

Woodpeckers send sound messages. They tap their bills against tree trunks. Different woodpeckers tap different patterns.

FACT FILE

Name: Great spotted woodpecker

Where it lives: forests

What it eats: insects and nuts

Type of animal: bird

Size: up to 9.4 inches (24 cm)

Body: black and white feathers with a red patch on the belly

Toes hold onto tree trunks and branches

Woodpeckers make their nests in tree trunks.

MINI FACTS

Woodpeckers peck trees up to 12,000 times a day.

Quiz

Test your skills!
Can you answer these questions?
The answers are on page 24.

1. What is the name for the low noises an elephant makes?

2. Where does a firefly make light?

3. What is a rattlesnake's rattle made from?

4. What is a baby skunk called?

Useful Words

infrasound Very low-pitched sound that humans cannot hear.

keratin A tough material. Scales, feathers, and hair are made from it.

mate An animal that has babies with another animal.

migrate To move from one place to another to have babies or to find food.

molt Shedding skin that is too small or to lose fur or feathers.

predator An animal that hunts other animals for food.

prey An animal that is hunted by other animals for food.

scent A smell that is produced by an animal that helps to identify it.

Find Out More

Books

How to Speak Animal, Gaby Wild and Aubre Andrus (National Geographic Kids, 2022)

How to Talk to a Tiger ... And Other Animals: How Critters Communicate in the Wild, Jason Bittel (Harry N. Abrams, 2021)

Websites

www.bbc.co.uk/bitesize/articles/zv3kxyc

www.natgeokids.com/uk/discover/animals/sea-life/humpback-song/

www.wonderopolis.org/wonder/how-do-animals-communicate

Index

infrasound 8

light 4, 10–11

rattle 16–17

scents 9, 12–13
smell 9, 13

sound 4–5, 7–9, 12, 20–21
spray 13, 18–19

touch 8

whale songs 6–7

Quiz Answers: 1. The low sounds are called infrasound. **2.** A firefly makes light inside its body. **3.** The rattle is made from keratin. **4.** A baby skunk is called a kit.